About the cover: Autumn scene with falling leaves is a phenomenon after the fall of man, and is discussed in *Does God Really Love Us?* and this book by the author, using the science derived in *Electrino Physics* by the author.

The *Physics of Genesis* Draft 3

by Gordon Ziegler

The Physics of Genesis
Draft 3

© 2007 Gordon L. Ziegler
© 2013 Gordon L. Ziegler
© 2014 Gordon L. Ziegler
© 2015 Gordon L. Ziegler

All rights reserved.

Last revised January 1, 2015

Unless otherwise noted,
all Scripture references are from the
King James Version (KJV) of the Bible.

Author:
Gordon L. Ziegler
P.O. Box 1162
Olympia, WA 98507-1162 USA
ben_ent100@comcast.net

Preface

I wrote *Electrino Physics* with as few references to God and the Bible as possible, to be in the physics tradition, and so as not to offend physicists. But for many years I dreamed of writing a book named *The Physics of Genesis*. I longed to apply the origin relating calculations I was making in the Electrino Fusion Model of Elementary Particles to the scientific discussion of the origin concepts in the Bible book of Genesis.

For 24 years this book simmered in the back of my mind. But here it is!—a third draft of it.

The book starts with a brief summary of the model in *Electrino Physics*, but then shows from the calculations the omnipresence, the immortality, and the extreme wisdom of the One or Ones that originated and sustained and sustains the Universe. From other calculations I show the extreme improbability of the origin of the Universe in a Big Bang of pair production of particles. I show the truth of the Bible regarding the nature of the smallest particles. Things that the ancient Bible writers could not know and understand from the then current physical theory are revealed in the Bible—showing the Bible to be the inspired Word of God. "In the beginning God created the heaven and the earth" is now not myth, fable, legend, or falsehood—it is good science.

Genesis not only tells of the creation of the Universe and this earth, it tells of the fall of man and the great changes that brought to this earth. The discovery of how to reverse the order to disorder arrow in the second law of thermodynamics, related in *Electrino Physics*, sheds light on what apparently happened in the transition from pre-fall to post-fall earth, why God allowed the changes, and how God designs to reverse the damage, now, through science (see *Does God Really Love Us?* by myself).

This book concludes with the question, "How do you see it now?" The reader is invited to reflect on which model of physics best fits the needs of humanity.

Contents

Chapter   Page

Preface 4
Contents 6

1. Getting Started 8
2. The Sustainer 13
   A. Introduction 13
   B. Significance of the Uniting Definition 14
   C. The Sustainer 14
3. Ex-nihilo Creation 16
   A. Introduction 16
   B. Speaking Suns into Existence 16
   C. Speaking Worlds into Existence 23
   D. God's Ideal for His Children 23
4. The Creation Method 26
5. The Improbability of the Big Bang 28
6. Things Which Do Not Appear 30
7. The Effects of the Fall 31
8. Reversing the Second Law of Thermodynamics 33
   A. Nasty Law 33
   B. Reversing the Second Law of Thermodynamics .. 33
9. Miraculous Effects of the Refresher 35
   Reverse aging for adults 35
   Resurrections from the dead 35
   Backing diseases out of existence 35
   Reversing all decay 35
   Reversing pollution 36
   "Raising up the foundations of many

  generations" Isaiah 58:12 36
  Reversing forest fires 36
  Reversing all calamities; 36
  Reversing all effects of war; 36
  Preventing all munitions from firing; 36
"Making wars to cease to the end of the earth." Psalm
  46:9. 36
  Removing sinful propensities from people, including
  criminals; 36
  Emptying prison houses; 36
  Making possible and efficient Clean Energy
  Sources. 36
  The blessings of the Refresher are endless. 36
10. Recreating the Earth 37
  A. Alternate Method 37
  B. The Theory of Everything 37
11. Bog Dating 38
  Introduction 39
  Calculations by the Initial Inaccurate Assumptions ... 40
  Comments 42
12. How Do You See It Now? 44

Chapter 1

Getting Started

The year was 1957. The date was October 4. The Soviet Union surprised the United States and the West by being the first to put an artificial satellite into space—Sputnik. That event precipitated fear and panic in some parts of the world, like the Philippines. People thought Russia was more powerful than the United States, and they feared Russian domination and persecution.

American mathematicians and physicists also were alarmed at the American lag behind Russian technology and science. The mathematicians and physicists did their best to alleviate the situation by redesigning texts for students. They wanted them not to get bogged down in needless memorization. The educators wanted to give a coherent, but concise ordered summary of physics research, and to elicit imagination, intuition, and creativity in science research, so the students would make discoveries in science. The Physical Science Study Committee were authors of *PSSC Physics*, a high school physics text, which was the first physics text studied by this present author. The author had an exceedingly inspiring math and physics teacher, Robert R. Ludeman, at Upper Columbia Academy, near Spokane, Washington USA. The author's brain was filled with the centuries-long contest between wave and particle in the nature of light, and the final dual synthesis of those concepts in the current science, quantum mechanics.

By the way, February 1, 1958 the United States launched its first satellite 'Explorer.' The Filipinos jumped for joy, they were so happy. But the physics quest of the millennium was already started in the events previously described.

The author studied *PSSC Physics* the school year 1963-1964. By April 1964, the author had an intuitive leap—that there was an aether after all (Einstein notwithstanding). This he based on several evidences. One group of evidences was the results of several experiments testing for Special Relativity (see *Electrino Physics*, Chapter 2). Another evidence was the next thing to be expected from the pattern established by the centuries of the wave-particle controversy, followed by the dual model of light and matter. But a more influential evidence was the Bible itself: Einstein taught that nothing could go faster than the speed of light. But Daniel 9 recorded a prayer that can be read in five minutes—Gabriel was said to start flying from heaven at the beginning of that prayer, arriving on earth to interrupt Daniel's prayer at the conclusion of the prayer. Heaven is not in this solar system. It must be a minimum of several light years away—maybe much more. At any rate, Gabriel had to fly much faster than the speed of light for the Daniel 9 account to be true.

Some have said, "but angels are spirits, not flesh and bone. Maybe things material like flesh and bone cannot travel faster than the speed of light." Jesus' own experience answers this objection. On the morning of the resurrection, Jesus ascended to the Father in heaven, but descended to meet with the disciples on the road to Emmaus and in the Upper Room that evening. He later explained that "a spirit hath not flesh and bones, as ye see me have." Luke 24:39. Jesus had flesh and bones even after his resurrection. It did not take Jesus years to fly to heaven and back. Jesus flew faster than the speed of light (Einstein notwithstanding). When some of his classmates and teachers were swayed by Einstein's popular Theory of Relativity, and his own stand was unpopular, the author never vacillated from his concept arrived at that April, 1964—that there was an aether after all, and Einstein's Theory of Relativity was partially in error. At the young age of 17 he aspired to be the one to derive relativity in an aether. He did not realize that Hendrick

Antoon Lorentz already derived it, even before Einstein's theory, and that in Lorentz's theory, the speed of light barrier was no problem. In 1977 the author derived relativity in an aether himself.

In 1969-1970 the author first wrote a summary of the centuries old wave-particle controversy over the nature of light in a religious text, *The Black Box of Time*. In 1990 he adapted that chapter as the beginning chapter of a precursor manuscript of *Electrino Physics*. That manuscript consolidated also *Relativity in an Aether*, 1977, and *Manipulating Gravity and Inertia in a New Model of the Universe*, 1982.

In 1990 the author realized it was not "relativity in an aether" precisely, but "quasi-relativity in an aether" (relativity up to a point). In the main, the principles of relativity and equivalence hold true, but in the absolute, both principles are false, which facts are demonstrated by the thought experiments propounded by the author in *Electrino Physics*, Chapter 2. In that chapter and Chapter 3, the author derives "Special Quasi-Relativity in an Aether" from first principles. The author shows that aether special relativity is in harmony with observations at typical angles of observation, but varies greatly from the expected in other configurations—and why that was never detected before.

Electrino Physics, Chapters 2 and 3, shows that the famous Lorentz Transformations can be derived from first principles in an aether system. The connection is stronger in this model to the Galilean Transformation than in Einstein's derivation. Special Quasi-Relativity in an Aether is in harmony with all the tests and observations relative to Special Relativity.

Electrino Physics, Chapter 4, shows that General Quasi-Relativity in an Aether derives easily from Special Quasi-Relativity in an Aether through putting the gravitational escape velocity transformation in the radius of the spherical coordinates of the Lorentz Line Element, arriving at the Schwarzschild Line Element. The Schwarzschild Line Element is employed in Einstein's General

Theory of Relativity. And from it most tests for General Relativity can be calculated. Numerically "General Quasi-Relativity in an Aether" is the same as "General Theory of Relativity," except that there is an exciting explanation of the 3 term in the equation for the perihelic shift of Mercury in the aether model. The 3 is not just the way the equations turn out, it shows that there is an orbital component to aether motion, which, by the way, was the very thing Michelson and Morley tried to measure in their watershed experiment.

Electrino Physics, Chapter 5, demonstrates that our failure to have a model of gravity and inertia is the fault of Einstein's aether-less Theory of Relativity. With an aether, deriving a model of them is straight forward—uniting them in one formula is easy. This chapter gives us hope that mankind may develop inertia-less and gravity-free craft.

In *Electrino Physics* Chapter 6, mathematical models of electrons and photons are calculated from first principles. Up until now, uniting Special and General Relativity in particle physics has been seen to be as difficult as uniting fire and ice. In the aether model, however, Special Quasi-Relativity in an Aether and General Quasi-Relativity in an Aether are exact fits in the model calculations. But that is only if fracton charges come in $\pm e$, $\pm e/2$, $\pm e/4$, and $\pm e/8$ (as in the Electrino Hypothesis), not in $\pm 2e/3$ and $\pm e/3$ (as in the Quark Hypothesis). In Appendix B of *Electrino Physics*, the structures are induced for every known particle. There is no need for quarks and gluons in this model. Appendix A gives an example how Appendix B structures can be utilized in a new method of balancing particle decays called chonomics—giving the chonomics for leptons.

Chapter 7 of *Electrino Physics* presents a Unified Field Theory which unites six forces and the weak force, which is the sum of an infinite number of interactions—which were identified and formalized in a paper, G. L. Ziegler, "A New Way to Calculate

Electron and Muon g/2-factors," *Galilean Electrodynamics*, Vol. 17, No.1, January/February 2006. An alternate method of calculating the same things is in Chapter 8 of *Electrino Physics*—calculated when the author doubted electrinos could orbit faster than the speed of light, which is theorized in Chapter 21. Both calculations may be useful for research. The unification in Chapter 7 comes largely in unification of a set of pairs of forces. To one person, there are "an obscene number of forces" in this model. But they are all united to one parent force in this model—gravity. They all play distinct roles which can be calculated. The author has spent years collecting these forces, distinguishing them from other similar forces.

Chapter 2

THE SUSTAINER

A. Introduction

In *Electrino Physics* Chapters 1-7, a model of physics—a Grand Unification Theory (GUT)—is developed based upon seven postulates and one ad hoc hypothesis (number 6). In Chapter 8 a sweeping unification is made of forces and systems with only two postulates—the parsimony principle and a formula predicting strong mass due to relativity. (See equation (2-1) below.) After the writing of that chapter, the author discovered he needed a third input into his system—the mass of the electron or the ratios of the radii in the relativistic inner frame and non-relativistic outer frames of particles. That cannot be derived, but must be input into the system. But, at any rate, one formula stood out in that chapter:

$$\frac{0\, arbitrary\, mass\, unit}{0} \equiv M_0(arb.ma.u.). \qquad (2\text{-}1)$$

For how fundamental that definition is to all the forces and systems in the Universe, read *Electrino* Physics or *Electrino Physics* Draft 2 or Draft 3, Chapter 8, Unifying the Universe! M_0 is the strong mass of a whole particle in the relativistic frame as seen by an observer at rest, 0 in the numerator on the left side of the equation is the mass of a whole particle (uniton) at rest, and 1/0 is the gamma factor transforming the non-relativistic frame to the relativistic frame when the uniton has exactly the speed of light relative to the aether. Many fundamental constants, forces, and particles in the Universe can be derived from this simple definition and the parsimony principle.

B.   Significance of the Uniting Definition

Since M_0 is in terms of $\hbar$, c, and G, the application of the definition shows that the speed of light c, the gravitational constant G, and Planck's constant over 2π $\hbar$ are not dependant on matter density in the Universe, temperature, or pressure, or any other thing. They only appear complex due to the units we measure them in— kg, m, s. In natural units they are all 1 and changeless. The Universe has an underlying simplicity to it.

A tremendous benefit of this discovery, also, is the great harmony this can bring to some of science. Much is hard derived from one uniting master definition.

C.   The Sustainer

A shockingly obvious result of this postulate foundation of our calculations is that the formula, otherwise indeterminate, must be defined. But who defines it, and how can it be dependably defined in all particles, in all regions of the Universe, in all ages of the Universe's history? This is a big job! It is an essentially important job. The very existence of the Universe depends on it. We cannot define it. It was defined long before us—long before the creation of the human race, and far beyond the confines of this earth.

Equation (2-1) gives a surprisingly complete blueprint of the One that sustains the Universe:
1) That One must be before all things.
2) By that One must all things consist.
3) That One must be eternal.
4) That One must be immortal.
5) That One must be omnipresent.
6) That One must be exceedingly wise.

7) That One must never change the plan of sustaining the Universe.

The Bible provides us the descriptions of One that fulfills all these scientific criteria of our needed Sustainer of the Universe. Compare the following numbered references with the numbered descriptions in the previous list:
1) "He is before all things." Colossians 1:17.
2) "By Him all things consist." Colossians 1:17.
3) "The King eternal . . . ." 1 Timothy 1:17.
4) "The King . . . immortal. . . ." 1 Timothy 1:17.
5) God is everywhere.   Psalm 139:7-12.
6) "The only wise God." 1 Timothy 1:17.
7) "For I am the LORD, I change not." Malachi 3:6; ". . . for that He is strong in power: not one [star] faileth." Isaiah 40:26.

The Bible identifies that One as God in three Persons—the Father, Son, and Holy Ghost (Matthew 28:19). Accident cannot define Eq. (2-1) for all particles, for all locations, and all time in the history of the Universe. But that is essential to our very existence! Our origin cannot be a series of accidents. Is it now remarkable that the Bible fulfills the deepest needs and the most up-to-date discoveries of science?

The stars in the galaxies burn under the second law of thermodynamics as we know it. They would collapse unless the Sustainer would also periodically range from star to star and refresh them by reversing the Hydrogen to Helium processes in each star by temporarily reversing the order to disorder arrow in the second law of thermodynamics in the star, and replenishing the lost Hydrogen in each star by ex-nihilo creation. See later chapters.

Chapter 3

Ex-nihilo Creation

A. Introduction

Theologians speak a lot about ex-nihilo (out of nothing) creation, which they suppose only God the Creator can do and creatures cannot do. But God is now giving humanity His secret of ex-nihilo creation. If we can see how to do ex-nihilo creation ourselves, certainly God could do it in the creation of our biosphere.

B. Speaking Suns into Existence

Gordon Lewis Ziegler had an extensive vision in a motel room in Salt Lake City, Utah on or about January 1, 1984 that lasted about an hour and a half. Among many other points, the vision pointed out Gordon's heavenly assignment of being the first to receive from God His trade secrets that set Him apart as God to be distinguished from His creatures. Gordon was to theorize and design a machine that could reverse adult aging, back diseases out of existence, and resurrect the dead. That one is about to be built, waiting momentarily funding. The second assignment was to receive from God the theory of how to speak stars and worlds into existence out of nothing, and to expand the Universe. That discussion we will begin in this chapter.

Gordon has long believed that creation particles, octons, can fuse to anti-quartons, which can fuse to semions, which can fuse to anti-unitons, which, with the corresponding charge conjugant particles, compose all the building blocks of light, gravitons, matter, and antimatter. Octons and anti-octons are creation particles—not part of gravitons, regular light, matter, or antimatter. A unified field theory and unified particle theory [1], [2] can be formulated without octons and anti-octons. The entire experience of humankind before has been without octons and anti-octons. They have never been discovered before by mortals.

Octons and anti-octons are so small and so outside of human experience that one way we can study them is to postulate their properties, build the Universe with them in theory, and then see if that theory corresponds to the particle properties we measure in the Universe. Octons, quartons, semions, and unitons are all electrinos; and their charge conjugates are all anti-electrinos. Like all electrinos and anti-electrinos, isolated octons and anti-octons should be spin less. The only way they would contribute spin is if they were in orbit around each other. One particle, octal light, does just that. That total spin of the octal light photon is 1 ℏ. That is the same as ordinary light photons. Octal light and light photons cannot be distinguished by their spins. But they can be distinguished be their energies. The energy of a light photon is E = hν; the energy of an octal light photon is E =⅛hν. This gives the experimenter a handle to detect octal light by. Another property of octal light is its photon has net zero mass, and therefore it travels precisely at the speed of light c—just the same as ordinary light.

Through the spin relation $mrc = n\hbar$, we see that if the spin of a particle is 0, the mass of the particle must also be 0. The masses of octons, quartons, semions, and unitons and their anti-particles are all 0. But that does not mean that quartons, semions, or unitons can ionize nothing like octons, for they are all compound particles. Compound particles cannot ionize nothing. But no wonder octons can ionize nothing—they have infinite charge to mass ratios, and they are not always bound in speed of light barriers. Since the mass of octons is 0, they all travel at ± c.

The radius of the octon is another surprise. If mass equals 0 and n equals 0, the spin relation mrc = nℏ could be defined to reduce to rc =ℏ or r = ℏ/c or anything. But it doesn't. The radius in the spin relation is defined to be ⅛ R_0. This calculates to be ≈ i 2.020 2484(12) x 10^{-36} meters.

And now, let us consider the actual ionization of nothing (not the aether particles, but nothing). What is the velocity of nothing? It can only be zero. The bare octon travels precisely at ±c when it ionizes nothing. The charged bare octon streaking at the speed of light through nothing is an action. According to Newton's Third Law, we have "for every action there is an equal and opposite reaction." To try to cancel out a positive traveling octon, nature produces a similar traveling negative octon through ionizing the nothing. But another positive octon is produced also. Without outside intervention, the negative octon and the initial positive octon just annihilate, leaving the new positive octon to replace the initial positive octon. We have changed nothing! Perhaps that is another safeguard against inadvertent explosive creation.

There are only two technologies or combinations thereof that can modify an ionization out of nothing for our purposes. The first is a magnetic field surrounding an ionization. But this makes no separation of the positive and negative octons. Without it, the octon-anti-octon pairs attract each other to annihilation.

The other is a pair of biased electric plates on the front and back of the ionization. The electrified plates will separate the differently charged bare octons and attract them to the oppositely charged electric plates. As the bare octons migrate to the charged plates they continually ionize more bare octons of both polarities, which begin new trails of charged octons to the charged plates and more ionizations. It is an explosive creation of octons separated in the main by the charged plates.

When the bare octons get to the electric plates, they pass through the plates without colliding or friction because the octons have zero spin and are bosons. Their charges, however, are attracted to the charged plates. The octons that pass through the plates reverse course because of the charge on the plates. They pass through the plates again, setting up an oscillation without friction. The charged plates are reservoirs of charged octons until the voltages are reversed on the plates and the octons are repulsed and ejected from the vicinity of the plates or the voltages are turned to zero.

Can electric plates, say one square meter, reservoir enough octons to build a star? Let us take our own sun as a typical star. How many atoms of Hydrogen are in our sun? About $1.2E57$. How many octons does it take to make an atom of Hydrogen? 32. How many octons would it take to make our sun? $3.8E58$. Not considering the boson character of octons and their cloud array about the plates, how many octons in a single layer would fit on a one meter square plate? $8.1E84$. That is a lot more than is needed to build a star, so we could have much smaller plates.

When an octon is not with an ionization, it is an action stirring up another ionization. The cycling time is fast.

If octons would collide with anti-octons head on at the relative velocity 2c, would they act as bosons and go through each other without colliding, would they annihilate each other, or would they bounce off each other? Also if octons would collide with octons head on at the relative velocity 2c, would they act as bosons and go through each other without colliding, would they fuse, or would they bounce off each other? There is probably not a single man in the world that could answer those questions for sure. But fortunately, we do not have to know everything to know how to fuse anti-octons into quartons, fuse quartons into anti-semions, and fuse anti-semions into unitons and speak suns into existence!

If head on octons and anti-octons barely missed each other, they would go into a slightly slower than the speed of light orbit of octal light photons held together by the meso-electric (ex-nihilo) force of opposites attracting. Their masses would cancel out to zero, therefore, like light, they would travel at the speed of light. They would be stable, and could go billions of light years without decaying.

If head on octons barely missed each other, they could go into a slightly faster than light orbit, held together by the strong electric force of like attracting like. These orbiting pairs of octons would have spins and magnetic fields, which could attract other orbiting octon pairs into close reactions between the orbiting pairs of octons.

If two pairs of octons were attracted into the same orbit in the same direction, the strong electric force would fuse the four octons into two orbiting anti-quartons. And two pairs of anti-quartons in like proximity would fuse into two orbiting semions. Ditto with two pairs of semions into two like anti-unitons, which would repel each other, and stop the chain fusion.

Whole particles cannot fuse. Like-charged unitons repel each other. Only fractons can fuse. Dissimilar particles cannot fuse. Only two like particles can fuse at a time. It is the speed of light barriers that put matter particles upon a firm foundation.

But if two positive quartons barely missed each other, they would induce an orbit of positive quartons. This is huge! Because orbiting quartons have positive mass, whereas straight line quartons have zero mass. We have created mass and energy out of nothing!

If our positive quarton near missed the other positive quarton on its right side, the quarton orbit or ring would have down spin. If they near missed on the other quarton's left side, the orbit would have up spin.

Newly fused quartons have a multi-stage journey to their stable state in pions or annihilation gammas: First of all quartons are attracted to each other by the strong electric force of like attracting like. They go in orbit about each other, and sort of form quarton orbital rings. These orbiting rings have spin and magnetic moments and strong electric force for each other. They have north and south poles axially. If one ring approached another quarton ring axially north to south north to south they would attract each other magnetically and attract each other with the strong electric force. But the quarton rings are not solid. They have space between the pairs of quartons. Therefore the two quarton rings attract each other into the same orbit, and we have four quartons going the same way in the same orbit. When that happens the strong force of like attracting like takes over, and the four positive quartons fuse into two positive orbiting anti-semions in a positive electron (positron), which will annihilate an ambient electron to make a $\gamma\gamma$ burst. This is not the path to pions.

The path to pions is if the quarton rings approach a NSSN or SNNS configuration—the rings then repel each other and flip over until the edges of the rings attract each other with a NS configuration. In this SN case the quartons in the rings orbit in the opposite directions from each other. But again, the rings are not solid. There is space between the quartons in the sort of rings. Thus the edges of the two rings attract each other, and the opposite orbiting quartons come into the same orbiting plane—with the quartons colliding with each other. They either eternally go through each other without colliding, or they eternally bounce off each other over and over again setting up an eternal stable chatter—but they do not fuse or annihilate each other. The spins will add to zero, the magnetic fields will cancel out, and the particle system becomes a net zero spin boson pion with mass equal to 139.57018(35) MeV.

But notice the masses of the quarton rings do not cancel out. They are not matter and antimatter. They are both matter. The masses add. The pion is one of three principal building blocks for any matter.

We already saw how to make semions through the fusion of anti-quartons. The -1c vector left over from the fusion process gives a second level speed of light barrier about the particle (a black hole within a black hole), as well as converts the particle from antimatter to matter. Now if we have a beam of semions barely miss each other, they will form electrons with $0.510998910 \pm 0.000000013$ MeV [3] mass. Again this is mass and energy out of nothing! Electrons and anti-electrons (positrons) are the second principal building blocks of matter.

We already saw how to make anti-unitons by fusing semions. Unitons and anti-unitons are encased in triple black holes. They are whole particles and cannot further fuse or un-fuse. They are stable particles in the Universe—building blocks of light and matter. In that, with the creation of matter, unitons continually increase, the expansion of the universe is assured. Though unitons are whole particles, they cannot abide alone. Only two particles so far have been observed to accompany unitons, + spin electrons for most particles and pions for Σ particles. When an electron is made to orbit a uniton, a neutron is produced. Neutrons have heavy masses (939.56536 ± 0.00008 MeV) [3] and are the third and final principal building blocks of matter that are whole particles and attach to other whole particles.

Knowing what fuses to what and how to make mass and energy out of nothing is only half the problem. We have to do it extremely rapidly to form a star. How can we do it? If we have an array of charged plates to steer particle beams around to fuse and orbit them, we will be limited to small results. We need a grand slam of particles, hopefully with adequate quantities of each particle formed. After some consideration, the author concludes that all the principal building blocks of matter would be formed in the operation of a single pair of plates before and behind an initial ionization. The collision of like charged particle beams would take place in the space surrounding the charged plates in the octon reservoir. Suppression of the final antimatter could be accomplished by having the positive charge on the positive plates an amount to be determined less than the negative charge on negative plates. The processes could be speeded up or slowed down by raising or lowering both positive and negative voltages. To stop the reactions, both voltages could be lowered to zero. To start the reactions, with voltages in operational range, ionize a captive octal light photon with an electron beam in the area between the plates. The resultant free octons will be bosons, which will not be contained in the trap, but will be controlled by the electric plates if the voltages are high enough.

The first creator device should contain a ten meter cubed cavity with one side open to the vacuum of space in the direction you want the new star. The device should be carried to deep space within an inertia less craft and made exterior to the craft at destination. The craft should have the proper injection velocity for the star at time of creation. Inside the open cavity should be two electric plates insulated from the cavity and connected to a battery operated, computer controlled, voice actuated system. For the purposes of the Everlasting Father God, Who inspired much of this work, a star should best be spoken into existence by a little child.

We cannot experiment with octons unless we have some octons to experiment with. They cannot be fissioned from quartons, semions, unitons, or their anti-particles. They cannot be obtained from regular light or matter. The only hope of obtaining octons on earth is to find an outside source of octal light shining on earth. That is not as hopeless as it sounds.

Stem cells are special miracle cells. But they are hard to distinguish from other cells. But the long search is worth it. That is the biological analogy to octal light photons. They are hard to find and differentiate from ordinary light photons. But the search is worth it! Octons can create suns and worlds out of nothing!

The octal light and octons are associated with creation and God. The Bible mentions the face of God shining on us. Maybe octal light comes from the face of God. Where is God? He has spent a considerable amount of time in heaven. Where is heaven? Prophets say in the open space in Orion. We should train our octal light detection equipment on a beam from the open space in Orion and catch an octal light photon in a trap.

Octal light streaming from afar and octal light trapped in a mirrored box are octons in the safe mode. As long as they are in that mode, they will not explosively ionize or create. Octal light cannot be ionized by any amount of octal light or regular light. Such beams just go through each other without colliding. Octal light, however, could be ionized by a small beam of electrons made portable with the trap small enough to easily fit in the cavity to initiate sequence.

Even without new worlds, this device, mass produced and operated throughout the Universe, could expand the Universe and save it from collapse on itself and destruction.

C. Speaking Worlds into Existence

Actually, speaking worlds into existence would be much more difficult and time consuming, and would require much more sophisticated equipment and planning than speaking suns into existence. The makeup of worlds would require many more different kinds of atoms and molecules than simple Hydrogen. Those atoms and molecules would have to be synthesized from protons, neutrons, and electrons produced by the equipment for the speaking of suns into existence, and deposited in layers or other configurations in the world-to-be. There would have to be detailed plans for the composition of the world-to-be, and fabrication by layers of similar materials would be the simplest method of building a world. Genesis does not describe such a process for our world. Instead God remodeled a pre-existing planet when creating the biosphere of our world.

D. God's Ideal for His Children

God's ideal for His children has years ago been prophesized to be the advancement of human technology to such a state that adults could play a new kind of Santa Clause hoax on innocent children. God's ideal for people would be to assemble equipment so vast, so cascading sophisticated, so hidden from view by space warp, so computer controlled by voice actuation, that parents could speak to each other and to a little child thus:

"Daddy, we really need another world badly! Why don't you see if you can speak a new world into existence?"

Daddy says, "Let there be a world!"

Nothing happens.

Daddy says, "I can't seem to do it today. Mommy, why don't you try?"

Mommy says, "Let there be a world!"

Nothing happens.

"Mommy says, "I can't seem to do it today either."

Then speaking to the little child, mommy says, "Honey, we need a new world very much! Can you try to say it, 'Let there be a world!'"

The little child lisps, "Let-there-be-a-world!"

And from that lisp, very rapidly, a dimly lit new world comes into existence!

The parents then remark, "You did it! Why don't you try to say, 'Let there be light!'"

The child lisps, "Let-there-be-light!"

Instantly the orb is bathed in a glorious light!

Then daddy says, "O thank you honey! This world will never cease to exist! It will endure to all eternity! Be careful what you long for, honey, for dreams do come true! There is nothing in this Universe that you cannot have if you want it badly enough, and you stick with it!"

Then the child will not fear and love a fictitious, judgmental, but benevolent saint. He or she will be in holy fear of himself/herself. He/she will resolve to only do good with his/her powers.

---

[1] Gordon L. Ziegler, *Electrino Physics*, Chapter 7 (PO Box 1162, Olympia, Washington USA 98507-1162: published by the author, 2011). http://benevolententerprises.org Book List.

[2] Gordon L. Ziegler and Iris I. Koch, *Prediction of the Masses of Every Known Particle (as of 2008), Step 2, Part 1,* Chapter 1 (PO

Box 1162, Olympia, Washington USA 98507-1162: published by the author, 2011). http://benevolententerprises.org Book List.

[3]   C. Amsler *et al.* (Particle Data Group), PL **B667**, 1 (2008) (URL: http://pdg.lbl.gov).

Chapter 4

The Creation Method

Electrino Physics, Chapter 6, also begins the consideration of the quantization of spin and the method of the creation of particles. Electrinos come in unitons, semions, quartons, octons, but not hexadecons, which would violate spin rules. The one way tendency of electrino fusion indicates that octons would fuse to quartons, which would fuse to semions, which would fuse to unitons, which would not fuse further. And electrinos in this Universe would not fission. At any rate, octons are the beginning particles.

Man has never controlled octons. All our control so far has been of particles containing quartons, semions, and unitons. But electrino physics indicates that octons should exist. We can see that octons are creation particles. All things can be made of them.

Man can walk or run, but he can travel faster if he makes automobiles, airplanes, or rockets. He can estimate time, but he can be more precise if he makes watches or atomic clocks. He can manipulate DNA in test tubes, but he can be more precise and rapid if he makes DNA sequencers and synthesizers. Just so, perhaps God can extend and accelerate His creative powers by making machines to sequence and synthesize octons in particles, and sequence and synthesize those particles in higher structures, like minerals, vegetation, animals, or people. Each desired object has its computerized mathematical template. Maybe that was the way God created the Universe and this earth. Or maybe His Spirit was so adept and rapid at creating, that He didn't need any machine. He could do it all Himself. But at any rate, God apparently used the synthesis of octons in His creation of things. When he wanted a bird, He synthesized a bird from octons. He was not indebted to pre-existing matter.

This method of creating is in contradistinction to the Big Bang Theory and Evolution. Instead of a Big Bang, God could create everything as needed—worlds, stars, galaxies. Also the original creatures could be made perfect to start with, not slow adaptations of genetics through mutations.

Incidentally, at the University of Washington, Seattle, Washington USA, the author observed a series of experiments to try to induce beneficial mutations in fish through the irradiation of fish eggs. The radiation caused mutations, but none beneficial. The more radiation the eggs were exposed to, the more stunted and deformed the hatched fish were. But there were no beneficial mutations. The evolutionary theory of the adaptation of species through beneficial mutations is fallacious.

Instead, now, electrino theory of physics points to intelligent design for the origin of this Universe. Now again, "In the beginning God created the heaven and the earth" (Genesis 1:1) is good—the best—science.

Chapter 5

The Improbability of the Big Bang

New to human science in *Electrino Physics* is the idea of elementary particle fusion—of electrino fusion. The idea was first brought out in Chapter 6 of *Electrino Physics*. Chapter 12 more fully brings out this phenomenon for artificially induced fusion. In that chapter is a section on the origin of the Universe, which is quoted in part below.

"The electrino fusion model of elementary particles may revise our model of the origin of the Universe. While an all matter Universe could be constructed using electrino fusion techniques, an all matter Universe could not result from a Big Bang of pair production. . . . A Big Bang should produce as much anti-matter as matter. There is no electrical asymmetry which should dictate that all positrons should be fused to protons and neutrons and not that electrons should be fused to anti-protons and anti-neutrons. . . . This begs for an obvious interpretation affecting our models of the origin of the Universe. . . ."

The author in *Electrino Physics* did not interpret this shockingly obvious clue as to the origin of the Universe. But the author will here connect the dots. We live in a Universe where there is more matter than antimatter. That is not possible through pair production alone, but it would be possible by fusing the anti-semions in positrons from pair production to the core particles of protons and neutrons, and not by fusing the semions in electrons to the core particles of anti-protons and anti-neutrons. But there is no natural law that positron anti-semions should be fused and not electron semions. This could not happen as an accident or natural law. It had to be the result of the deliberate choice of one or more

Super Intelligent and Super Powerful Benevolent Beings acting in concert by convention.

Chapter 6

Things Which Do Not Appear

"Through faith we understand that the worlds were framed by the word of God, so that things which are seen were not made of things which do appear." Hebrews 11:3.

According to Electrino Fusion Model of Elementary Particles, by the author, everything in the Universe is made of electrinos. But electrinos do not appear. They cannot be made to appear with any degree of enlargement in a microscope. They do not appear in an electron microscope. Why? Because electrinos all have imaginary radii (the radii have the square root of minus one in them). Things which do appear have real radii. According to Hebrews 11:3, the things which are seen were not made of things which do appear. Electrinos do not appear. Hebrews 11:3 is in harmony with the Electrino Hypothesis. Is it remarkable that the Bible should be in harmony with what in some ways is the most advanced model of physics? How could the writer of Hebrews know that? Could he learn it from any then current model of science? No. He could only know it through revelation by the Spirit of God. The Bible is in harmony with the most advanced, most up to date model of physics.

Chapter 7

The Effects of the Fall

The Bible makes plain several effects of the fall of man. There was a sense of nakedness after the fall that there was not before the fall. Genesis 3:7. Adam and Eve were fearful of the approach of God after the fall, which they were not before the fall. Genesis 3:8-10. The soil was more difficult to till and work after the fall. After the fall there were thorns and thistles, which there were not before the fall. Genesis 3:17-19. After the fall, people, animals, and vegetation returned to dust after their deaths. Genesis 3:19. That phenomenon didn't exist before the fall. Then there was no death of man or beast. These are all effects of what we now know as the order-to-disorder arrow in the second law of thermodynamics. In the world order we live in after the Genesis fall, order tends to disorder—people and things grow old, sick, and die. But that was not the way it was in the beginning. Up until now, we have not known a way around it. But the author discovered a way around that deleterious law in *Electrino Physics*, Chapter 16. In the book, *Does God Really Love Us?* the author enlarges upon that scientific plan, and shows that it is one of the Biblical prophetic scenarios for the last days. God wants us to work together and do it. One person can not do it. But the human race can do it—even a small subset of it.

The effects on Adam and Eve and their posterity and the earth after the fall were not only caused by the removal of the Tree of Life from the diet of man. Genesis 3:22. They were a change in the force fields outside the Garden of Eden, as compared with the force fields inside the Garden of Eden. But now we may know the difference in those force fields. It is opposite directions for the order to disorder arrow in the two regions. It would be like a toggle switch in the second law of thermodynamics. God switched that toggle in the area outside Eden by merely removing a process in that area that kept all things healthy and alive.

But why would God do this [switch the toggle in the second law of thermodynamics to the deleterious phase] to the human race? What would have happened had God not removed that protection from the human race? The negative effects and consequences of the violation of God's law would quickly be backed out of existence, but there would be no ability to shed blood to make an atonement for sin. Also the human race would have no just sense of the enormity of sin and its consequences. Adam and Eve and the human race needed to complete the experiment with sin and find out just how bad it is. The Messiah needed to come and die and atone for sins. But apparently God held in reserve the reversal of the order-to-disorder arrow in the second law of thermodynamics for the last days, as an effective mechanism for restoring order on this planet and restoring all the inhabitants to health, happiness, and holiness. But the post-fall changes are correctly attributed to contingency plans by God, not to accident or automatic consequences of the sin. Genesis 3:22-24. This point is in harmony with our science model.

Chapter 8

Reversing the Second Law of Thermodynamics

A. Nasty law

There is a law in physics (the second law of thermodynamics) that in a closed system all reactions and energy transfers must tend from more order to more disorder. This law, in most cases, is a nasty law. It is the law which makes people grow old, sick, and die, and their bodies to disintegrate to dust. It is the law which makes it easier to do bad things than good things. Mankind has been exposed to this law ever since the fall of Adam and Eve in the Garden of Eden in the beginning of the world. Before that, there was just the reverse of that law. Everything tended from more disorder to more order. People, animals, plants, and things did not grow old, sick, or die. It was easier to do good things than bad things.

B. **Reversing the Second Law of Thermodynomics**

Recently the author discovered how to reverse the second law of thermodynamics from the post-fall state to the pre-fall state using the fusion of positron anti-semions. Originally, and in heaven, God apparently did it through the fusion of octons. That process is instantaneous—"in the twinkling of an eye." 1 Corinthians 15:50-54. The process through the fusion of positron anti-semions is slower and safer for the wicked. The method God used originally through octons puts up with no evil. While it is life to the righteous, it is death to the wicked.

How can we reverse the second law of thermodynamics using positrons? The harmful second law of thermodynamics we now experience is when the arrow between order and disorder points toward the disorder. It is when the change of order energy over time is negative or zero≤ 0). The beneficial reversed second law of thermodynamics is when the arrow between order and disorder points toward more order. It is when the change in order energy over time is positive (>0). (Order energy is just the positive or negative energy in the creation or annihilation of particles.) It is naturally negative, but we can make it slightly positive by fusing the anti-semions in positrons to unitons (the core particles of protons and neutrons).

It doesn't take much of that to do the trick—10 trillionths of an Amp beam of axial oriented positrons accelerated to 940 MeV in an accelerator and collided with 10 trillionths of an Amp beam inverted axial oriented positrons accelerated to 940 MeV in another accelerator would reverse the order to disorder arrow to nearly five acres of land. The effect is backwards to what we would expect. Higher beam currents result in smaller areas affected, and lower beam currents collided result in wider areas affected.

Chapter 9

Miraculous Effects of the Refresher

Reverse aging for adults

The simplest effect of the Refresher to understand is reversing adult aging. Old people can be made young adults again in the active footprint of the Refresher. This effect for positron anti-semion fusion does not really back up time or the clock. It merely reverses the order to disorder arrow in adults. It saturates at the maximum state of order—which is young adulthood. It reverses adult aging at a rate of about 1836 times as fast as the rate the original adult aging occurred. A century of aging can be reversed in just under 20 days.

Resurrections from the dead

The reverse aging occurs also for bodily remains—re-assembling dust and bones into living beings again. All the dead of all ages of earth's history would be resurrected in about 3½ years of Refresher machine time, starting with those who died most recently.

Backing diseases out of existence

In the process of reverse aging, diseases would be backed out of existence. This would work also for difficult diseases like HIV AIDS, cancer, cystic fibrosis, and Ebola.

Reversing all decay

Spoiled fruit would un-spoil in the active footprint of the Refresher. Fresh fruit would stay at the maximum state of order for fruit forever—fresh picked fruit. And this would be without refrigeration. This would amount to a new kind of food preservation without canning or freezing or drying or salting.

This process would un-decay everything in the Refresher footprint, not just fruit. And the footprint could be enlarged to cover the entire earth.

Reversing pollution out of existence
In the Refresher footprint, all pollution would be reversed out of existence. Depending on the Refresher control settings, this effect could be world-wide.

"Raising up the foundations of many generations" Isaiah 58:12.
The Refresher would automatically rebuild previous decayed structures. It would rebuild and restore the entire earth.

Reversing forest fires
The Refresher not only would stop forest fires in its footprint, but would reverse the fires—restoring all that was lost—animate and inanimate, including lost trees and homes.

Reversing all calamities;
Reversing all effects of war;
Preventing all munitions from firing;
"Making wars to cease to the end of the earth." Psalm 46:9.
Removing sinful propensities from people, including criminals;
Emptying prison houses;
Making possible and efficient Clean Energy Sources.
The blessings of the Refresher are endless. In short it would restore earth to Edenic perfection in about 3½ years of machine time.

Chapter 10

Recreating the Earth

A. Alternate Method

Eden lost hungers for Eden restored. Genesis predicts the Serpent shall be trodden under foot of the Seed of the woman (Genesis 3:15; Romans 12:20). The first dominion shall come to the Tower of the flock (Micah 4:8). God's people long thought that would come through the creative acts of Jesus Christ after His Second Coming. But the Great Controversy is in overtime, and Jehovah Immanuel has taken the field. The earth itself may now be restored in a different way—through His technology of reversing the order to disorder arrow in the second law of thermodynamics globally. A machine of His design may now be built to restore all things in preparation for the Coming of our great Prince—the Lord Jesus Christ. One machine, the Refresher 1, could make old people young again, back diseases out of existence, un-decay everything, resurrect everything, and one step at a time restore the earth to Edenic perfection. The entire process would take about three and a half years machine time.

B. The Theory of Everything

The theory of reversing the order to disorder arrow in the Second Law of Thermodynamics is presented in *Refresher 1 Manual*, Chapter 2, and *Rad Waste-free Power*, Chapter 2, by the Author. This is but a portion of the aether theories of relativity and the Grand Unification Theory set forth in *Electrino Physics* Draft 2, *Advanced Electrino Physics* Draft 3, and *Predicting the Masses*, Volumes 1-3. The Standard Model of Physics is thought to be a Theory of Everything. But The Electrino Fusion Model of Elementary Particles predicts more things than the Standard Model, such as gravitons.

Chapter 11

Bog Dating

Analyses of Uranium 234 and Radium 226 values of average Uranium bogs show the topography of North America is less than 4,700 years old.

By G. L. Ziegler, P.O. Box 1162, Olympia, WA 98507-1162 USA, E-mail ben_ent100@msn.com.

Abstract

The age of a Uranium bog can be determined by the relative abundance of the daughter products of Uranium 234 in the bog. The author was made aware of the fact in the early 1980s during a state licensing of a Uranium bog mining company. More accurate than individual bog data are the federal averages on the activity levels of Radium 226 in Uranium bogs nationwide—12 picocuries per gram (pCi/g). The Uranium 234 levels can be calculated from published values, and the age of the bogs can be modeled. The result is the age of the bogs nationwide which is less than 4,700 years. This is considerably less than recent datings of the ice age.

Introduction

The significance of this paper is that the radioactive dating of Uranium bogs shows them to be much younger than current dating of the ice age.

In the work of Gordon L. Ziegler as Radiation Health Physicist for the State of Washington years ago, he stumbled across some data which might date Uranium bogs nationwide. His calculations put the ice age at less than 4,700 years ago, as opposed to 18,000 to 60,000 years ago as held commonly. [1,2]

At the Washington State Office of Radiation Control Section, Joseph S. Stohr interested the author in the dating calculations. Initial assumptions for calculating the age of the bogs were:

Comparing the activity of the daughter products to the parent Uranium is a way of dating the deposit. Thorium is much more insoluble than Uranium, so most of the Thorium may have been left behind when the Uranium migrated to the bog. This phenomenon approximately resets the nuclear clock to zero, because Thorium 230 is a daughter of Uranium 234 which in turn is a distant daughter of Uranium 238. Radium 226 is a daughter of Thorium 230. The Radium 226 half-life is only 1602 years. [3] (p. 112). Most Radium 226 that washed down with the Uranium would be decayed by now. The current Radium 226 would have to come from ingrown Thorium 230. Since the half-life of Thorium 230 is 8.40×10^4 y, or 50 times that of Radium 226, the Radium 226 would keep up well with the in growing Thorium 230 in a transient equilibrium. So the Radium 226 would be the same as the Thorium 230 activity.

Thorium 234, Palladium 234m and Palladium 234 have very short half-lives. So they would be in secular equilibrium with the Uranium 238. Since Uranium 234 and Uranium 238 are the same element, they have the same chemical and solubility properties. Therefore the Uranium 234 would not be separated from the Uranium 238 in the migration process. Thus the clock for Uranium 234 would not be set back to zero like the Thorium 230. The Uranium 234 would be old—in secular equilibrium with the Uranium 238. Therefore all we have to be concerned with is the in growth of Thorium 230 into Uranium 234, as the Radium 226 will follow the Thorium 230.

To illustrate the above decay chain, we present it in graphical form in Chart 1.

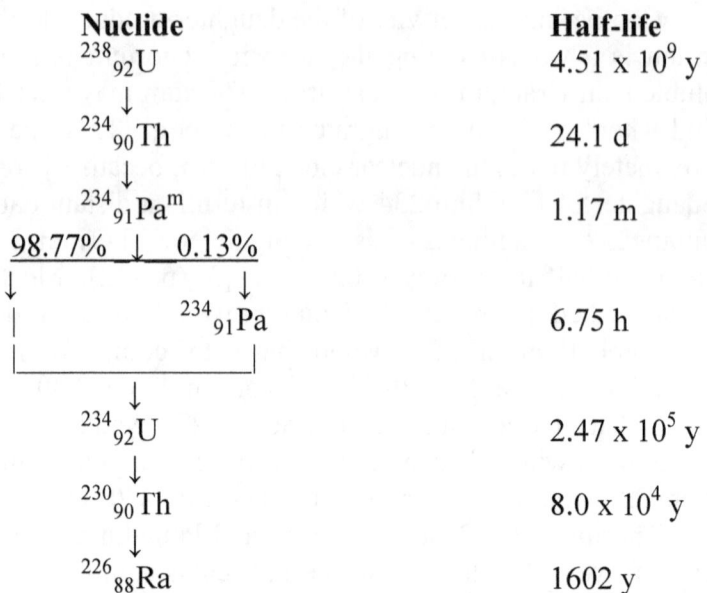

Chart 1: (Taken from [3].) This chart illustrates the decay scheme relationships between Uranium 238 and its daughter products down to Radium 226 together with the respective half-lives.

There are several inaccuracies in the initial assumptions. [a] Some Radium 226 undoubtedly washed down with the Uranium 234, and has not all decayed away. This would make the true age of the bog younger than calculated by the initial assumptions. [b] While Radium 226 activity would be the same as Thorium 230 activity after 100,000 years, unfortunately this is not true a few thousand years after a clock reset. In the early years, the Radium 226 would be high compared to the Thorium 230. This would make the true age of the bog younger than calculated by the initial assumptions. We will first calculate according to the initial inaccurate assumptions, then comment.

Calculations by the Initial Inaccurate Assumptions

Average data were taken from many Uranium bogs across the United States. Averages for a post glacial Uranium deposit mill are: ore grade U_3O_8—0.10%; Radium 226—12 pCi/g. [4] (p. 22).

Two more data points are needed to calculate the date of the Uranium bog. The two are provided by the *Radiological Health Handbook*: U234—6.19 x 10^{-3} Ci/g—0.0057% abundance. [3](p. 370.) The pCi/gram of the Uranium 234 in soil in the typical post glacial Uranium bog calculates from the above data:

$$\frac{0.0010 \text{g} U_3O_8}{\text{g bog soil}} \frac{0.848 \text{gU}}{\text{gU}_3O_8} \frac{0.000057 \text{gU}^{234}}{\text{gU}} \frac{6.19 \times 10^{-3} \text{Ci}}{\text{gU}^{234}} \frac{10^{12} \text{pCi}}{\text{Ci}} \approx \frac{300 \text{pCiU}^{234}}{\text{g bog soil}}$$

There are 12 pCi/g Radium 226 in bog soil. By the above inaccurate assumptions, there would be 12 pCi/g Thorium 230. That is 4.0% in growth (0.040).

To date the migration, one may calculate the time of in growth, using Equation 8-44 in Ralph T. Overman and Herbert M. Clark, *Radioisotope Techniques*. [5]

$$A_B = (A_A)_0 \, (1 - e^{-0.693/T_B}).$$

This equation is for secular equilibrium where half-life of parent is say 1000 times greater than that of the daughter, which is not the case for Thorium 230 in growth into Uranium 234 activity. But small in growth should make it an acceptable approximation. Note: This is not a close approximation. Because the half-life of the parent is not 1000 times that of the daughter, the activity of the parent is less than we would expect according to the assumption. Relatively speaking, the activity of the daughter is greater than we would expect. The in growth is faster in shorter time. The true age of the bog would be younger than calculated using this assumption. But we can still calculate the upper limit of the age of the bog using this assumption.

Solving for t we get

$$t = \frac{T_B}{-0.693} \ln\left[1 - \frac{A_B}{(A_A)_0}\right].$$

For Thorium 230 in growth of 4.0% into Uranium 234, we get:

$$t = \frac{8.0 \times 10^4}{-0.693} \ln[1 - 0.040] \approx 4,700 \, years.$$

Comments

Our radio-dating is a little on the long side, but that is to be expected. If any Radium 226 washed down with the Uranium 234, over one eighth of it would still be here. The in growth of Radium 226 and Thorium 230 would be somewhat less than we above calculated, and the in growth time would be less. Though Thorium is much less soluble than Uranium, some of it could still have been washed down with the Uranium. The in growth period would again be shorter. The above error would also make the real age of the bogs and the wholesale reconfiguring of earth's land masses more recent than our rough simple calculation.

In conclusion, the Uranium 234 and Radium 226 abundance data in bog soil are typical of Uranium bogs anywhere in North America. The dating then is of the topology of the earth's surface. This study shows that the world-wide topology altering event took place less than 4,700 years prior to 1960 to 1983, when the data were taken. This is much more recent than even the ice age by other dating means. The ice age was simply not as old as it has been thought to be.

References and Notes

[1] Christopher R. Scotese, Last Glacial Maximum, The Last Ice Age, Paleomap Project, http://www.scotese.com/lastice.html.

[2] The Last Ice Age, Climate Timeline Tool: Summary of 100000 Years, http://www.ngdc.noaa.gov/paleo/cyl/100k.html.

[3] *Radiological Health Handbook*, Revised Edition (Rockville, Maryland 20852: Bureau of Radiological Health and the Training Institute, Environmental Control Administration, U.S. Department of Health, Education, and Welfare, Public Health Service, Consumer Protection and Environmental Service, January 1970).

[4] Joseph S. Stohr and John L. Erickson, "Regulation of a Post-Glacial Uranium Deposit in the State of Washington," *Sixth Symposium on Uranium Mill Tailings Management*, Fort Collins, Colorado, February 1-3, 1984, Geotechnical Engineering Program, Civil Engineering Department, Colorado State University, Appendix A, Comparative Analysis.

[5] Ralph T. Overman and Herbert M. Clark, *Radioisotope Techniques* (New York: McGraw Hill Book Company, Inc., 1960) p. 305.

Acknowledgment: The initial help of Joseph S. Stohr is acknowledged. He did the lion's share of the work.

Chapter 12

How Do You See It Now?

New information has been presented to the reader in this book and *Electrino Physics*, by the author. Is it scientific? Is it true? Does it meet the needs of mankind? Current theories of physics have enough information to destroy the earth in a nuclear holocaust. The new model by the author has enough information to restore the entire earth to Edenic perfection. The current model of physics has enough information to kill you. The new theory of physics has enough information to heal you of any disease and make you live forever (provided the technology is funded). An origin by accident with no guarantee of a future life can be a depressing thought, but creation by intelligent design can give identity and purpose to life. True, it also gives accountability. God chastises and disciplines those He loves, but He is not willing that **any** should perish. 2 Peter 3:9. "For this is good and acceptable in the sight of God our Savior; Who **will have** all men to be saved, and to come unto the knowledge of the truth." 1 Timothy 2:3-4 KJV. Through the technology introduced in these books, and in *Refresher 1 Manual*, by the author, God could fulfill His purpose to save every spec of life in the history of the Universe. That includes you. God loves you. He is not willing for you to perish.

So how do you see it now? You have seen atheistic models of science, and now a model of science compatible with the Bible. You have seen a great dichotomy between accepted science and the Bible. Now you have seen a model of science that squares with the Bible. In the past you have had your opinions and taken your stands. But how do you see it now?

www.ingramcontent.com/pod-product-compliance
Lightning Source LLC
Chambersburg PA
CBHW051824170526
45167CB00005B/2144